KT-224-403

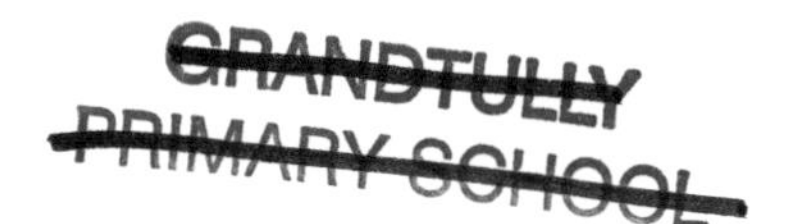
GRANDTULLY
PRIMARY SCHOOL

 MACHINE TECHNOLOGY 

Mixing Machines

Amanda Earl and Danielle Sensier

Titles in the series:

Cutting Machines

Digging Machines

Mixing Machines

Spinning Machines

Cover inset: This baker is putting the ingredients to make bread into a dough mixer. Find out how a dough mixer works on pages 18 – 19.

Title page: This boy is using a hand-held rotary whisk to mix up the ingredients to make pancakes (see page 9).

Series and book editor: Geraldine Purcell
Series designer: Helen White
Series consultant: Barbara Shepherd, (former) LEA adviser on the Design and Technology National Curriculum.
Photo stylist: Zoë Hargreaves

First published in 1994 by Wayland (Publishers) Limited
61 Western Road, Hove, East Sussex BN3 1JD, England.

British Library Cataloguing in Publication Data
Earl, Amanda
Mixing Machines. – (Machine Technology Series)
I. Title II. Sensier, Danielle
III. Bull, Peter IV. Series
621.8

ISBN 0 7502 1279 9

DTP design by White Design
Printed and bound L.E.G.O. S.p.A., Vicenza, Italy.

Words in **bold** appear in the glossary on page 30.

Contents

Mixing paint

These children are painting a picture of some fruit, but they do not have the right colour for the oranges. They will need to mix two different coloured paints, red and yellow, together to make orange-coloured paint.▼

▲ Mixing paint would be a very messy job if you had to do it with your hands. To make the job easier, you can use a hand tool, such as a paintbrush. This is a simple mixing machine.

When two different coloured paints are just poured into one dish they mix together slightly, but the colour is patchy. To make a better mixture, the paints need to be stirred. By stirring them together, the two colours become completely blended until they are one.

Making a mixture

To make a milkshake, two different **ingredients** are mixed together in the glass – a powder and a liquid. The design of a fork is good for mixing liquids because it has metal prongs. ▼

◀ When you move the fork around the glass, the prongs **agitate** the liquid until the powder is completely mixed.

▲ A spoon is also a good tool for making a mixture, such as coleslaw. This is a food made from salad cream and plain yoghurt with fresh, raw vegetables, such as grated carrots and shredded cabbage.

The salad cream and yoghurt are smooth and the vegetable pieces are rough, but the stirring action of the spoon helps to turn the different ingredients into one mixture.

Whisking food

When you make up a packet of instant dessert mix, the dessert powder and the milk should be whisked together. Whisking blends the two ingredients together and air is trapped into the mixture. The air makes the dessert light and fluffy.

A balloon whisk is a good tool for this job. As the whisk scrapes the bowl and is lifted in a circular movement, the mixture is agitated by the metal loops of the whisk. When this action is repeated, air bubbles are trapped in the mixture.▼

Pancake mixture is made by beating milk, flour and eggs together. ▼

Rotary whisks use **gears** to mix ingredients faster than a balloon whisk. The handle turns the crown wheel, the large wheel with double-sided teeth along its edge. These interlock with the teeth of the bevel gears on each side. The gears rotate and turn the beaters. ▶

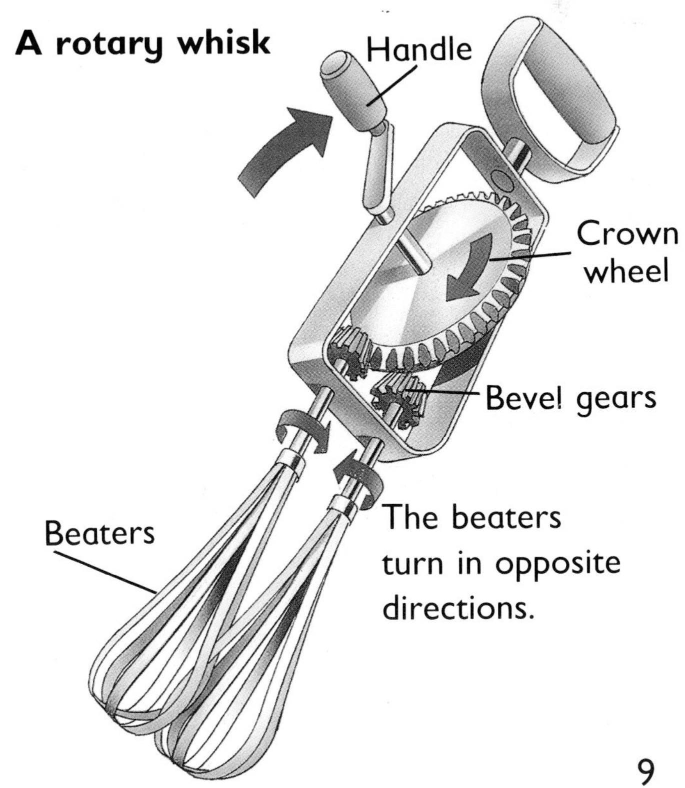

Churning butter

▲ In the past, butter used to be made by hand with the help of butter churns. Butter moulds, like the one in the photograph, were used to shape the butter into blocks.

A butter churn was a container with a beater, called a paddle, inside. The paddle was worked by turning a handle. Cream was placed in the churn. The handle was turned to move the paddle round to mix the cream. By mixing the cream over and over again, it separated into butter and buttermilk. ▼

A glass butter churn

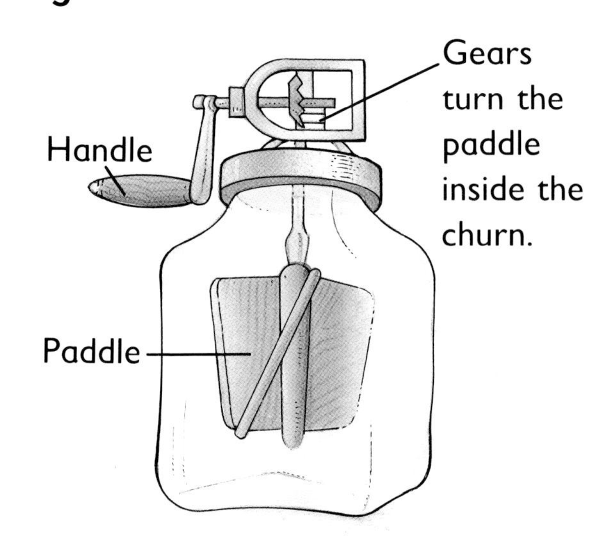

Gears turn the paddle inside the churn.
Handle
Paddle

To make larger amounts of butter, a bigger machine was needed, called a barrel churn. Some cream was poured into the barrel and the lid closed tight. When the handle on the barrel was turned, it flipped over. As it was flipped, the cream was shaken backwards and forwards to separate the butter. ▶

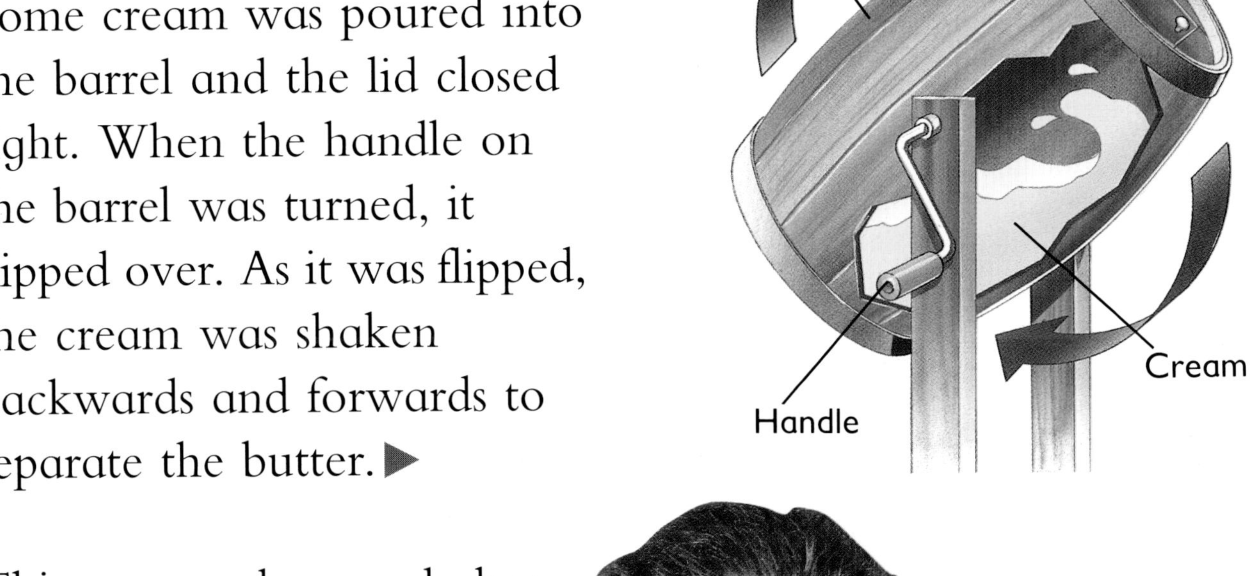

This woman has made her own butter in a home-made metal barrel churn. ▶

Shaking cocktails

Fruit cocktails are refreshing to drink and fun to make. They are made from freshly-squeezed fruit juices mixed with syrups or cream. The easiest way to make them is in a special container, called a cocktail shaker. When you shake a cocktail shaker up and down, the different liquids are mixed to make a single mixture.▼

▲ The design of a cocktail shaker helps you to mix drinks quickly, without any mess. Try stirring liquids together in a bowl – it is difficult to do this without them spilling over the sides. A cocktail shaker is a bit like a bottle with a screw-on top. You can shake the ingredients without spilling a drop!

Electric food mixer

A hand-held **electric** food mixer does the same job as a rotary whisk, but it works faster and takes less human effort. This is because it is worked by a small electric motor. The motor drives a worm gear, which is a rod with a screw thread round its length. The worm gear rotates two bevel gears and these turn two beaters at high speed. The beaters can break down lumps quickly and mix ingredients together into one mixture.▼

The inside of an electric food mixer

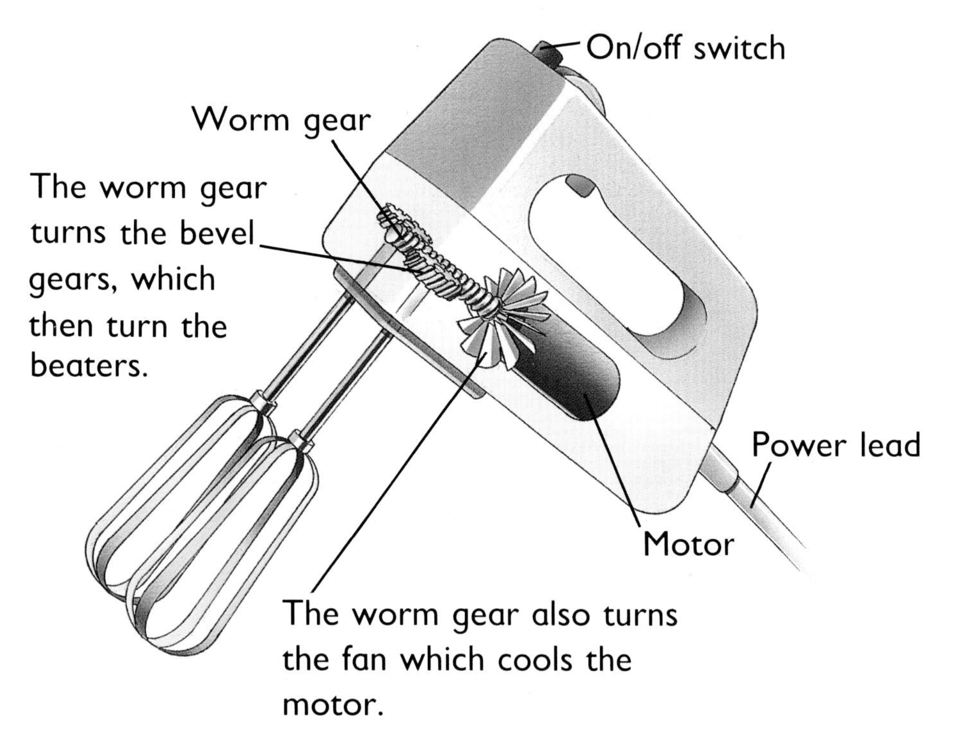

On/off switch
Worm gear
The worm gear turns the bevel gears, which then turn the beaters.
Power lead
Motor
The worm gear also turns the fan which cools the motor.

Electricity is dangerous. Ask an adult to help you use an electric food mixer.

▲ This girl is using an electric food mixer to make the cookie recipe below.

Peanut Butter Cookies

175g butter
1 egg (size 3)
50g crunchy peanut butter
275g plain flour
200g soft brown sugar

Ovens are hot and can be dangerous. Ask an adult to help you.

1. Put the butter, sugar and egg into a bowl and mix until smooth.
2. Add the flour and peanut butter. Mix together into a firm dough using the electric mixer on a slow setting.
3. Shape the dough into small balls and flatten slightly.
4. Place on a greased baking tray and bake in an oven for 10 – 12 minutes at 190°C (Gas Mark 5).

Food processor

The mixture to make a fruit cake has lots of ingredients. Sometimes it is too difficult to use a hand-held electric mixer for a job like this, so a more powerful mixing machine, called a food processor, is needed.

Food processors have different tools for doing different jobs. For mixing, they have a single beater that fits on to a central **spindle.** ▼

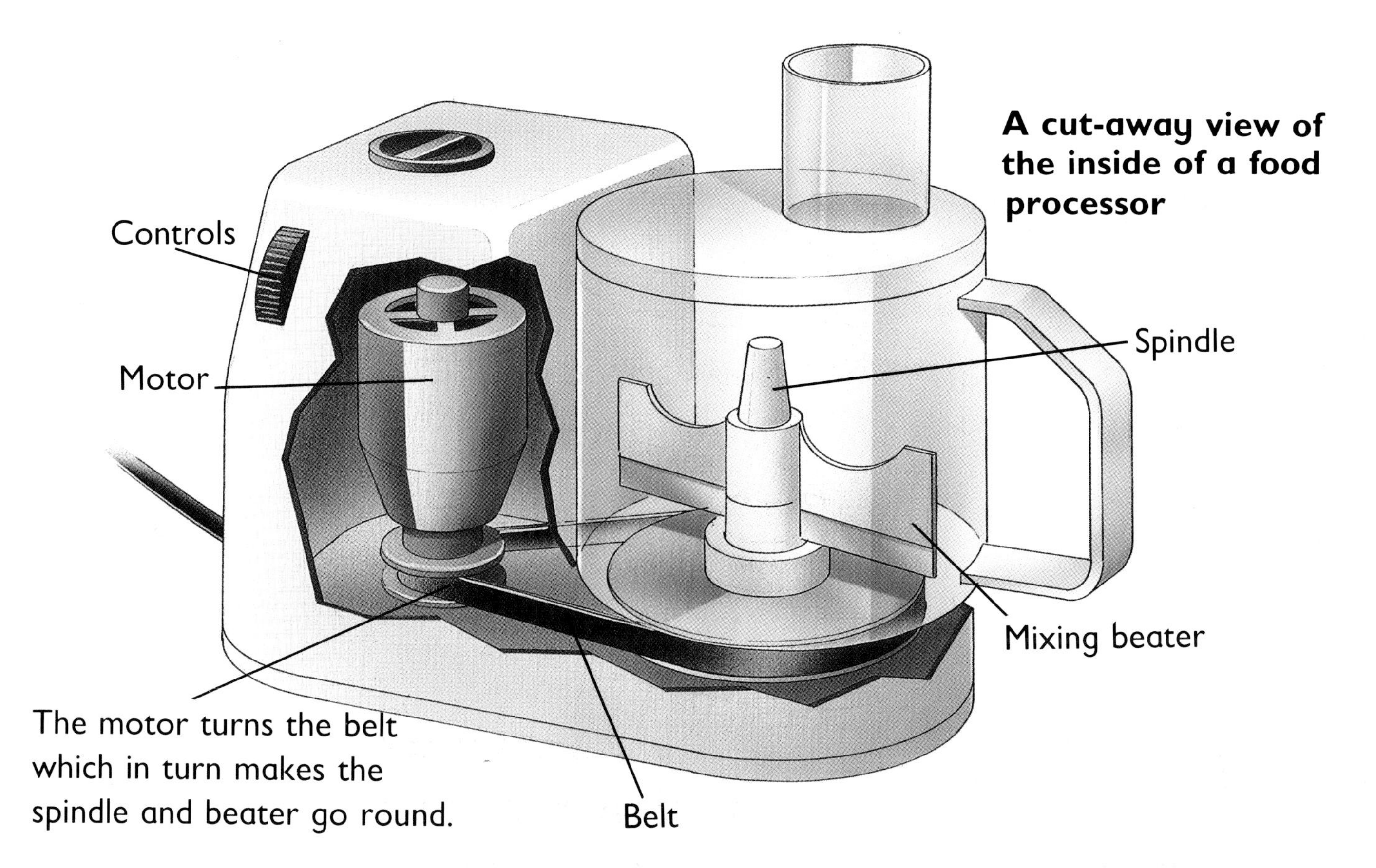

▲ The mixing beater turns round inside the bowl at high speed until the ingredients are mixed together. The beater is driven by an electric motor, which is connected to the spindle by a toothed belt.

In a hotel kitchen a food processor is used to mix bulky ingredients, such as sliced vegetables for soup. Because it is used so often the processor's bowl is made of hard-wearing stainless steel. ▶

Mixing dough

Bread is made from dough which is a mixture of flour, water, yeast, fat and salt. These have to be **kneaded** together thoroughly to mix the ingredients.▼

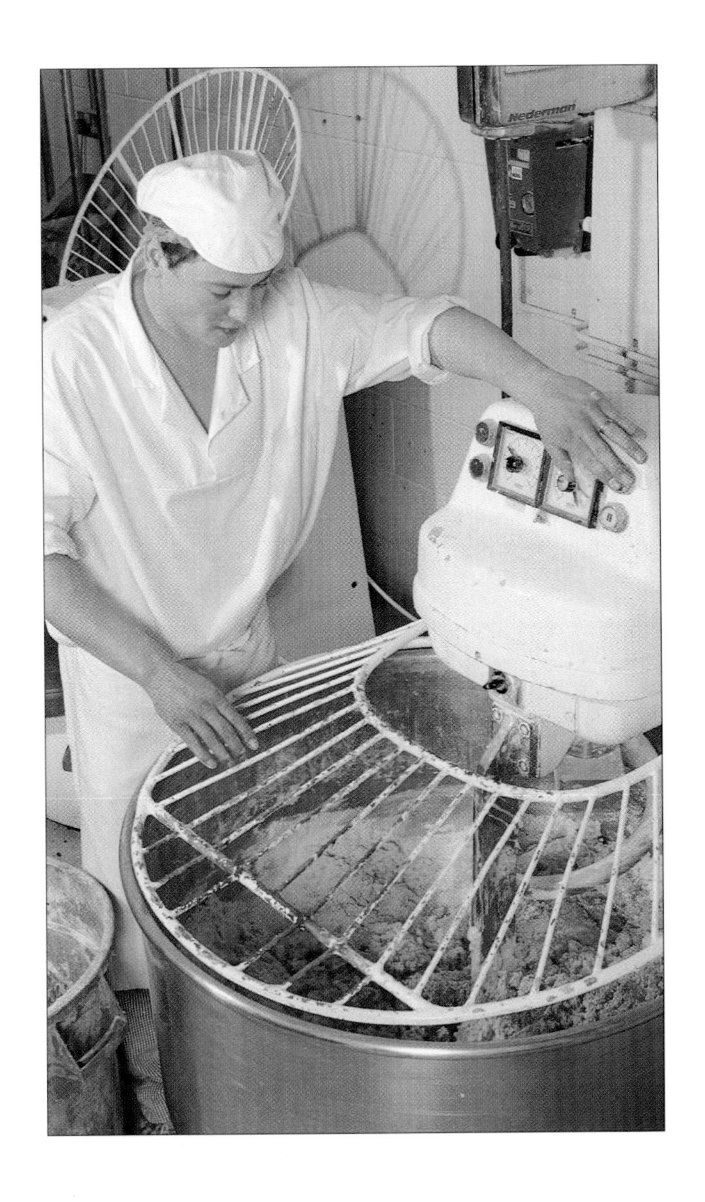

A cut-away view of a dough mixer

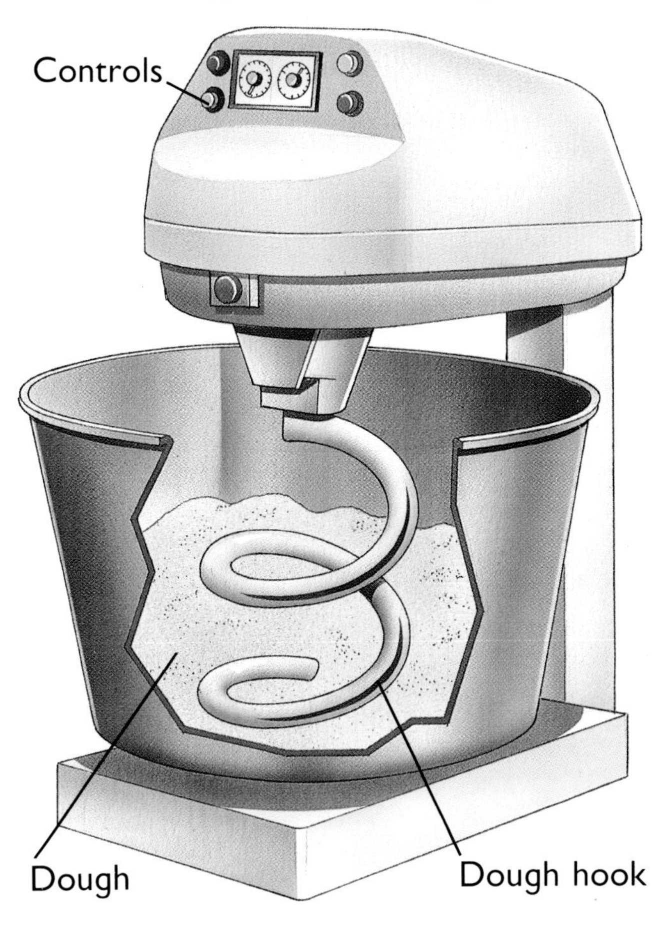

▲ In a bakery, bread dough has to be mixed in a machine called a dough mixer, as it would take far too long to mix by hand. The dough mixer has a very large bowl with a special tool, called a dough hook, in the centre. The hook is spiral shaped, which can mix the ingredients and cut through the thick dough at the same time. An ordinary beater would get stuck in the thick, sticky dough.

Mixing cement

Have you ever looked at a brick wall and noticed the layers of hard material between the bricks? This is called mortar, which is a mixture of sand, water and cement powder. The cement mortar sticks the bricks together.

Cement can be mixed with a spade, but it is a slow and messy job. It is much quicker to use a cement mixer. This mixing machine has a large metal drum which is turned round by a motor. The sand, water and cement powder are put into the drum and **baffle plates** inside mix them together as the drum turns.▼

▲ Once the mixture is ready, it can be emptied out easily by turning a small hand-wheel, which tips up the drum.

L.P.G.
MITCHELL
PLANT
14

Mixing concrete

Concrete is a building material made from a mixture of small stones, sand, cement powder and water. Once it dries, it forms a hard, solid material used to make the **foundations** of buildings and roads.

◀ A concrete mixer is a mobile machine, which can deliver ready-mixed concrete to a building site. It is a large lorry with a **revolving** drum on its back. A motor keeps the drum turning all the time, to stop the concrete from drying and going hard inside.

Inside the drum is a metal spiral, called an **Archimedes' screw**. As the drum turns, the spiral also turns to keep the mixture moving.▼

A cut-away view of the inside of a concrete mixer

Making paper

If you look at the pages of this book, it is difficult to imagine that the paper was once a soggy mixture, called pulp. Paper pulp is made from wood chips, waste paper, water and chemicals.

▲ A hydropulper is a giant mixing machine which is used in the paper-making process. This machine turns the pulp into a porridge-like mixture, called stock, by adding lots of hot water to the pulp and mixing it with huge paddles.

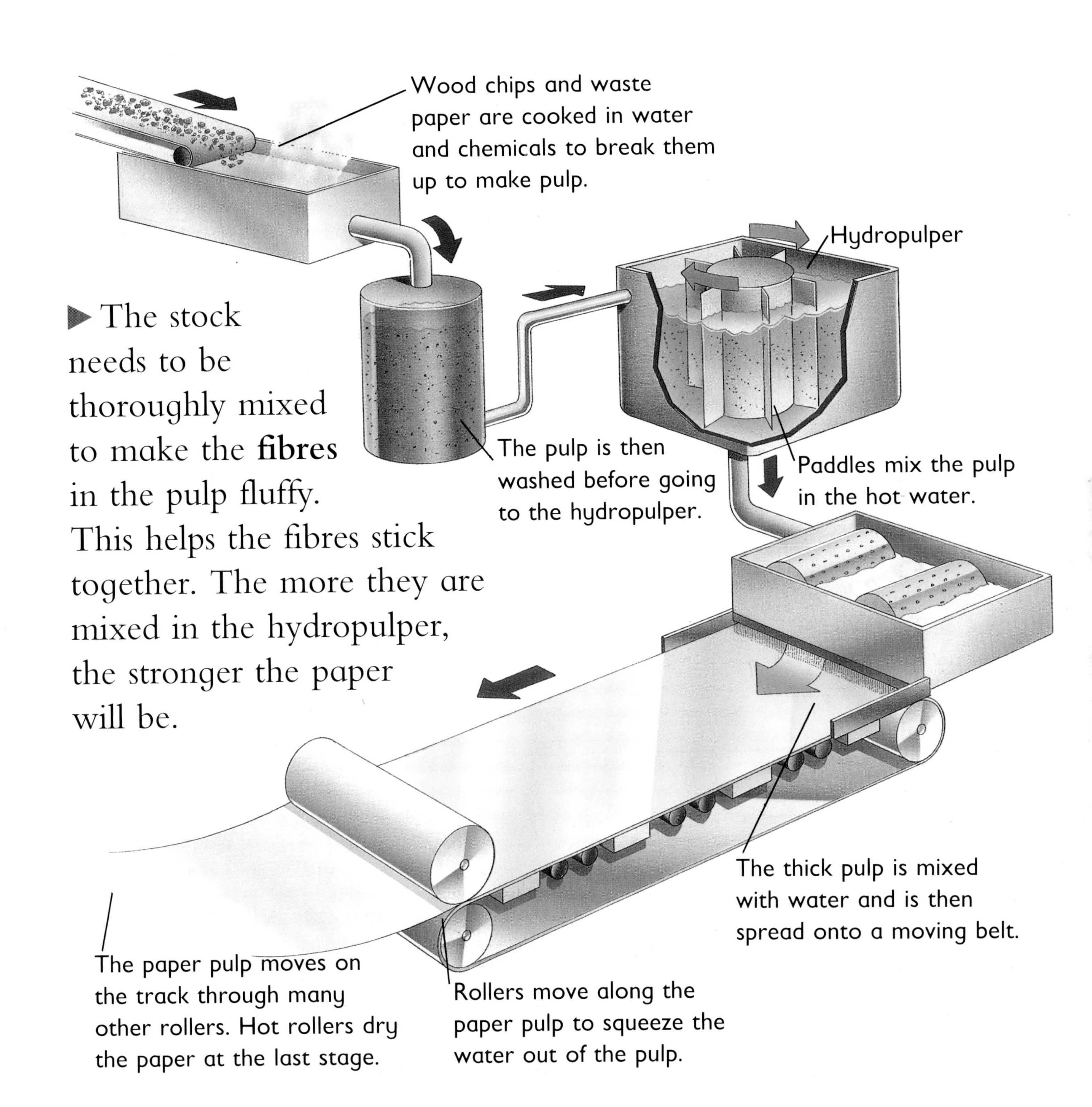

Wood chips and waste paper are cooked in water and chemicals to break them up to make pulp.
Hydropulper
The pulp is then washed before going to the hydropulper.
Paddles mix the pulp in the hot water.
The thick pulp is mixed with water and is then spread onto a moving belt.
The paper pulp moves on the track through many other rollers. Hot rollers dry the paper at the last stage.
Rollers move along the paper pulp to squeeze the water out of the pulp.

▶ The stock needs to be thoroughly mixed to make the **fibres** in the pulp fluffy. This helps the fibres stick together. The more they are mixed in the hydropulper, the stronger the paper will be.

▲ Other machines drain, flatten and dry the pulp mixture to make it into paper.

Blending chocolate

Milk chocolate is made from cocoa beans, sugar, cocoa butter and milk powder. These ingredients are mixed in large machines, called **grinding**, mixing, and **conching** machines.

▼ The grinding machine grinds the cocoa beans into a mixture, called cocoa mass.

The mixing machine mixes this with sugar and milk. Huge rotating paddles in the machine turn the mixture into a stiff paste. This paste is rolled into a smooth mixture, until it is ready for conching. ▼

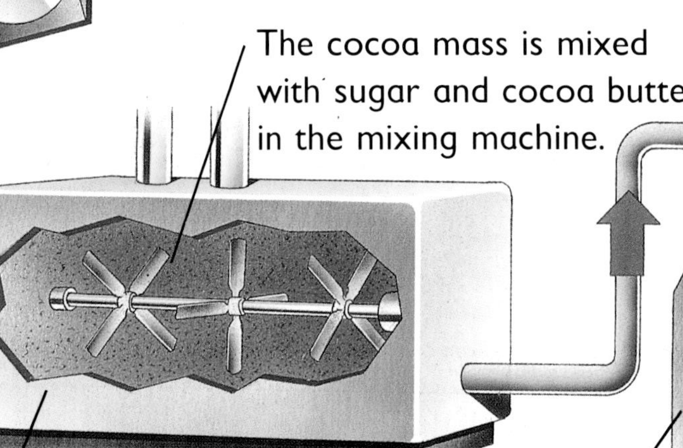

Roasted (cooked) cocoa beans go into the grinding machine.
Rollers in the grinding machine break up the cocoa beans to make cocoa mass.
Grinding machine
The cocoa mass is mixed with sugar and cocoa butter in the mixing machine.
Milk is added and the chocolate is stirred for up to three days in the conching machine.
Mixing machine
Conching machine

◀ The conching machine stirs the cocoa paste together with cocoa butter at high temperatures – up to 60–70°C. Rollers inside the conching machine move backwards and forwards, making large waves in the chocolate mixture. As the waves fold over, air is trapped in the mixture. This helps to make the flavour of the chocolate better.

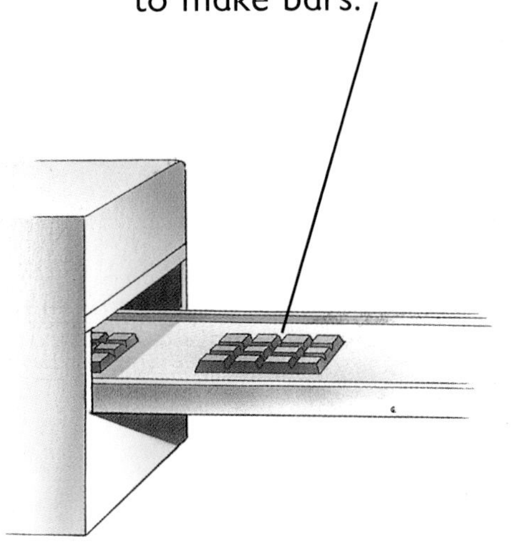

The liquid chocolate is then passed into moulds to make bars.

Making paint

Paint factories use grinding machines to make large amounts of different coloured paints. Coloured powders, called pigments, are mixed together with a liquid, such as oil (for gloss paint) or water (for emulsion paint). ▼

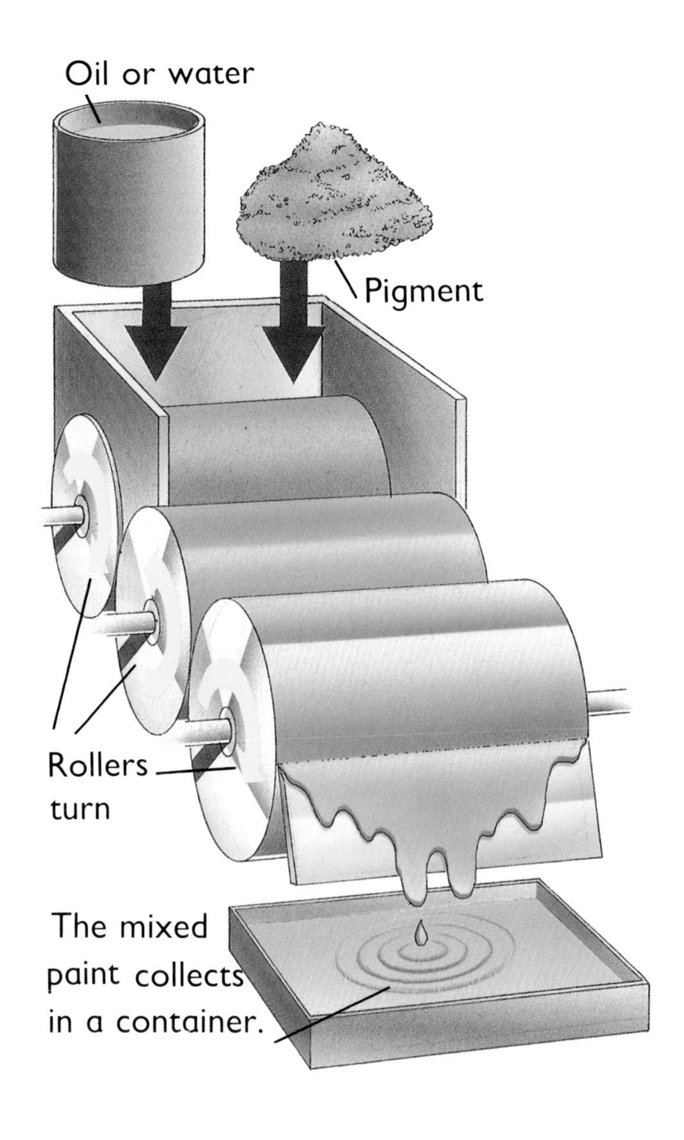

▲ A paint grinding machine forces the paint mixture over several large rollers. This rolling movement is better than a simple stirring action, because it breaks down the powder and mixes it more evenly throughout the oil or water.

Small paint mixing machines can be found in do-it-yourself stores. A tiny amount of liquid pigment is added to a tin of white paint to make the right colour, using controls at the top of the machine. This is placed in the bottom part of the machine. It is shaken at high speed until the colour is mixed evenly through the paint.▼

Glossary

agitate To move something quickly by shaking or stirring.

Archimedes' screw A spiral-shaped device which can move and lift water or liquid concrete. The liquid moves upwards along the spiral as it turns.

baffle plates Fixed metal sections in a cement mixer which are placed at angles to each other, like a cross. The cement is mixed by flowing round each plate as the drum revolves.

conching The process of stirring a liquid, such as chocolate, until it becomes a smooth mixture.

electric Relating to electricity, which is a form of energy. Electric machines have motors which change this energy into movement.

fibres Tiny threads that make up a material, such as paper.

foundations The part of a building which is built under ground. The foundations stop the building from sinking into the ground or from falling over.

gears Wheels with teeth around the edge that interlock with each other to change the speed or direction of a moving part of a tool or machine.

grinding Crushing something into a powder.

ingredients The different substances that go into a mixture.

kneaded To press and squeeze something together.

revolving To turn round and round.

spindle A pin or rod which turns round to move a tool fixed on to it.

Books to read

Chocolate (Food series)
by Jacqueline Dineen
(Wayland, 1990)
The Kingfisher Book of How Things Work by Steve Parker
(Kingfisher, 1991)
Machines (First Technology)
John Williams
(Wayland, 1993)
Paper by Sally Morgan
and Pauline Lalor
(Simon & Schuster, 1992)

Picture acknowledgements

Robert Harding Picture Library 21 (I. Robinson/Operation Raleigh); Tony Stone Worldwide 10 (M.Garrett), 22 - 3, 28; Topham Picture Library 11; Wayland Picture Library *cover background and inset, title page*, 4, 5, 6, 7, 8, 9, 12, 13, 15, 17, 18, 19, 20 (both) (APM Studios/Photo stylist Zoë Hargreaves)/ 16, 24 (P.Bennett/ SCA Euroliner), 27 (Cadbury Ltd). All artwork is by Peter Bull, except for the artwork on page 10 which is by John Yates.

Index